Living Things

by Sara E. Turner

Contents

The World Around You

Rocks

Flowers

Imagine you are in this field. Use your senses to observe. You might see rocks. You might smell flowers. And you might hear birds chirping. How are these things alike and different? This is a question that a scientist might ask.

Birds

Chapter 1

Big Idea Question

How Are Living and Nonliving Things Different?

Birds, such as puffins, are **living** things. Living things are alive. Things that are alive can move on their own and grow and change.

Living Things

Living things have **basic needs.** Most living things must have air, water, food, and a place or space in which to live.

Water to drink is a basic need for horses.

Living things must have these basic needs to grow and change. Without them, living things would die.

This tiny sunflower plant must have its basic needs met to grow into a tall sunflower plant.

Living things need other living things to stay alive. Humans and some animals eat plants for food.

Pandas eat bamboo and other plants.

Some humans and some animals eat other animals for food, too.

Kingfishers eat fish.

Nonliving Things

Rocks, soil, water, air, and light are **nonliving** things. They are part of nature. Boats, and other things that humans create, are nonliving things, too.

Water, boats, and sand are nonliving things.

Nonliving things don't need air, water, food, or a place in which to live. They don't grow. But some nonliving things can change over time.

Rocks change into sand over time.

Comparing Living and Nonliving Things

Living and nonliving things are different. You can compare a butterfly to a rock.

Butterfly

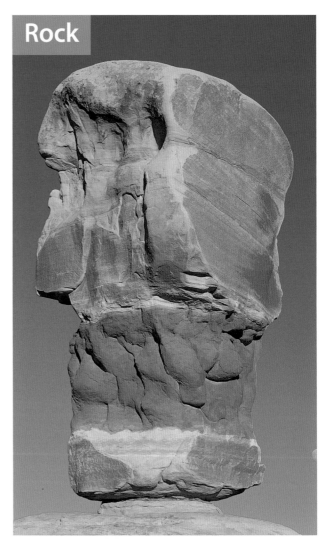

Rock

Living and Nonliving Things

Living Thing: Butterfly	Nonliving Thing: Rock
A butterfly must have air.	A rock doesn't need air.
A butterfly must have food and water.	A rock doesn't need food or water.
A butterfly moves on its own.	A rock can't move on its own.
A butterfly removes waste from its body.	A rock doesn't make waste.
A butterfly grows.	A rock can't grow.
A butterfly responds to the world around it.	A rock doesn't respond to the world around it.
A butterfly can make more butterflies like itself.	A rock can't make more rocks like itself.

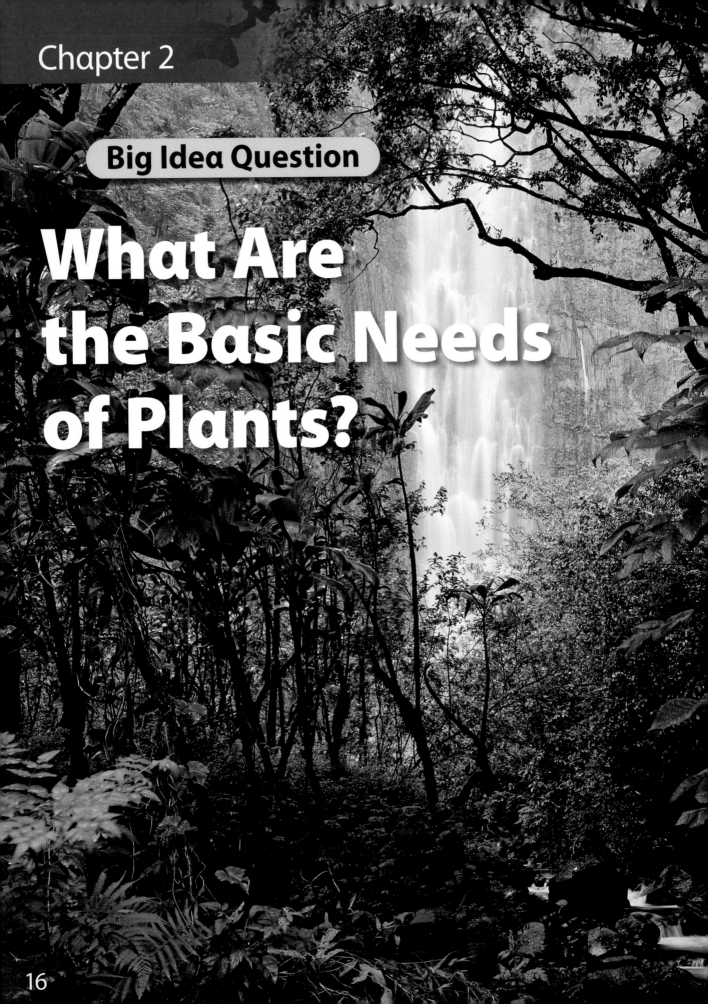

Chapter 2

Big Idea Question

What Are the Basic Needs of Plants?

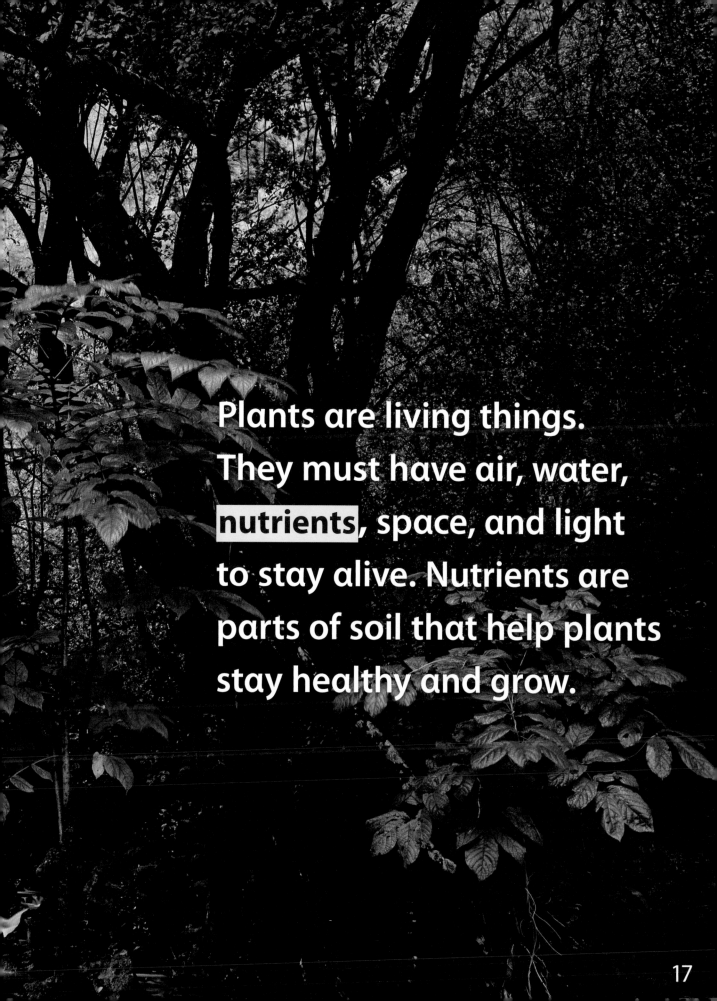

Plants are living things. They must have air, water, **nutrients**, space, and light to stay alive. Nutrients are parts of soil that help plants stay healthy and grow.

Parts of Plants

Plants have different parts. They use their parts, such as leaves, to get what they must have. All the parts work together to help a plant stay alive.

Needle-like leaves

Wide leaves

Leaves come in many sizes and shapes.

flower

leaf

stem

roots

19

What Plants Must Have

Plants must have food to grow and stay alive. Most plants make their own food.

How a Plant Makes Food

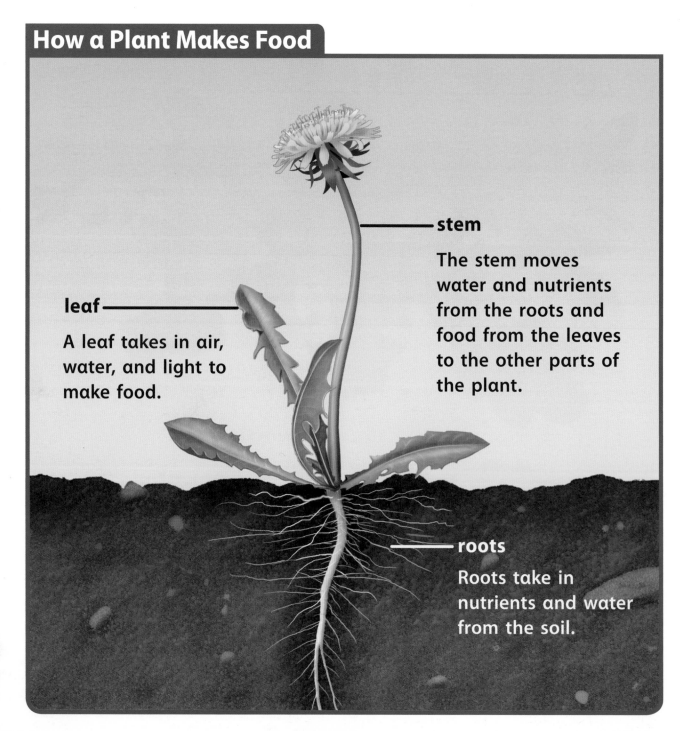

leaf

A leaf takes in air, water, and light to make food.

stem

The stem moves water and nutrients from the roots and food from the leaves to the other parts of the plant.

roots

Roots take in nutrients and water from the soil.

Plants also must have space to grow strong and to stay healthy.

This dandelion grows in a small space.

The plants in this field have a lot of space to grow.

Plants must have water to grow and stay healthy. What happens to a plant when it doesn't get water for a very long time?

This cactus plant has gotten the water it needs.

Look at the photos. The cactus plant on the left is healthy. The cactus plant on the right is not. How do the cactus plants look different?

This cactus plant has not gotten the water it needs.

Chapter 3

Big Idea Question

What Are the Basic Needs of Humans and Animals?

Humans and animals must have air, water, food, and a place or space to stay alive.

Air

Humans and many animals must have air to stay alive. Air is a nonliving thing. Different animals have different ways to get air depending on where they live on Earth.

This lemur lives in the rain forest. It breathes air into its lungs using its nose and mouth.

gill

Fish live in water and must have water to breathe. Gills help them breathe underwater.

Water

Like plants, humans and animals must have water to stay alive. Water is a nonliving thing. Humans drink water.

This girl is drinking water. She must have water to stay alive.

Animals must have water for drinking, too. Some animals, such as elephants, also use water to keep cool.

This elephant sprays itself with water to keep cool.

Food

What else must humans and animals have to stay alive? Food! Food is a basic need. Humans and animals get nutrients from the food they eat.

This boy is eating corn that was grown on a farm. Corn has nutrients.

This chipmunk is eating nuts that it found in a forest.

Most plants make their own food.
Humans and animals can't produce,
or make, their own food. They must
find plants and other animals to eat.

This squirrel eats a lot of food in the fall.
In the winter, food might be hard to find.

Shelter

Humans must have **shelter**. A shelter is a place to live. A shelter can give protection from the weather. Some shelters are made from plants. Others are made from metal and stone.

This shelter is made from tree trunks. It's a log cabin.

Some animals also must have shelter. A shelter can be a safe place for an animal to hide. It can also be a place where an animal can live and grow.

This elf owl made a nest in a cactus.

How Animals Stay Alive

Body parts, such as teeth, help animals get what they need to stay alive. Plant eaters, such as zebras, have different teeth than meat eaters, such as alligators. How are their teeth different?

This zebra uses its flat teeth to chew grass.

This alligator uses its sharp teeth to tear its food.

Animals also use their senses to stay alive. Cheetahs use their good eyesight to look for animals to eat.

This cheetah looks for other animals to eat.

Living Things Have Basic Needs

Living things must have certain things to stay alive and grow. Nonliving things do not have basic needs. They are not alive.

Plants are living things. They must have air, water, nutrients, space, and light to stay alive.

Humans and animals are living things, too. Humans and animals must have air, water, food, and shelter or space.

Glossary

basic needs (page 8)
Basic needs are what living things must have to stay alive.

Water is one of the **basic needs** horses have**.**

living (page 7)
Living things are alive.

A puffin is a **living** thing.

nonliving (page 12)
Nonliving things are not alive and never were.

Rocks are **nonliving** things.

nutrients (page 17)

Nutrients are parts of food and soil.

A plant's roots take in **nutrients** and water from soil.

shelter (page 32)

A **shelter** is a safe place where a living thing can make its home and grow.

This cactus is a **shelter** for this owl.

Index

National Geographic School Publishing
Hampton-Brown
www.NGSP.com

Printed in the USA.
Stromberg Allen, Tinley Park, IL

ISBN: 978-0-7362-5507-3

11 12 13 14 15 16 17

10 9 8 7 6 5 4 3